HOW TO FIND INFORMATION

•

MEDICINE AND BIOLOGY

By RUPERT LEE

Sponsored by Pfizer

Pfizer is a research-based, global healthcare company with headquarters in New York and sales in more than 150 countries.

Pfizer Ltd., located at Sandwich, on the Kent coast, is the largest Pfizer subsidiary in the United Kingdom. Some 2850 people are employed in an integrated business which involves the research, manufacture and selling of medicines for the treatment of disease in humans and animals.

 Working with Education

How to find information: medicine and biology.

ISBN 0-7123-0837-7

Published by:
The British Library, Science Reference and Information Service, 1996

© 1996 The British Library Board

Cover photograph: Dave Griffiths.

For more information on British Library publications contact Tony Antoniou at the Science Reference and Information Service, 25 Southampton Buildings, London WC2A 1AW. Tel: 0171-412 7471

DTP by Concerto, Leighton Buzzard, Bedfordshire (01525 378757)

Printed by PMC Printers, PMC House, 286–288 Brockley Road, London SE4 2RA

Contents

Acknowledgements iv

Author's note iv

1. Introduction. What's The Problem? 1

    The quantity of biomedical literature
    How to be sure of finding the relevant literature

2. Reference Sources. 2

    Encyclopaedias
    Pharmacopoeias
    Dictionaries

3. Electronic Searching 4

    CD-ROM and online, including concept searching and the use of Boolean logic in combining search terms

4. Manual Searching 15

    Introduction to manual versions of the major secondary sources

5. If the Library Doesn't Have It.... 19

    Where to go if your library fails you: union lists, inter-library loans, SRIS and the DSC

Acknowledgements

The author would like to express his thanks to Mrs. Susan Gove, librarian of St. George's Hospital Medical School, Tooting, London; and Ms. Magda Czigany, librarian of Imperial College, London, for their highly constructive comments on the first draft of the text.

Author's Note

How to Find Information: Medicine and Biology is intended to replace *How to Find Information: Life Sciences*, published in 1992 by The British Library and sponsored by Pfizer. It covers the same subject area as the earlier book, but has been largely re-written to account for recent innovations in information-searching techniques.

Chapter 1. Introduction – What's the Problem?

'Write an essay on Alzheimer's Disease. Here are some references'. 'Lecture on Diseases of the Central Nervous System: Reading List'. Reading lists of references drawn from scientific journals on a given topic are the core material of all science courses from second-year undergraduate level upwards. To most students, they are simply things produced by lecturers and tutors, and no-one asks how. The usual guess is that lecturers spend lots of time in the library, keeping themselves up-to-date on the subject that they teach.

But there comes a time (usually quite early on) in every scientist's career when it is necessary to learn about a particular subject from scratch, with nobody to point you in the right direction to start with. Suddenly you have to have a good up-to-date knowledge of, let's say, Kreuzfeld-Jakob Disease, which nobody in your particular laboratory or hospital has thought about much, and doesn't feature very largely in your textbooks (which are all a few years old, anyway). You go into the library... and then what do you do? The place is bulging with journals, textbooks and monographs, and among them somewhere are papers on Kreuzfeld-Jakob Disease. But how do you find them?

The best move is usually to consult your librarians. After all, they are there to help you to use the library. They can show you how to use not only the catalogues, but also the various other information sources such as abstracting journals, encyclopaedias, indexing services etc. They will also instruct you in the use of CD-ROM and networked databases. They may also be able to perform an 'online search' of the various bibliographic databases not available on CD-ROM. This can be particularly useful if there is a specialist database on your subject that is not available on CD-ROM.

But you will probably prefer to do the job yourself. After all, you know exactly what you are looking for, so you are the best person to find it. And modern CD-ROM's and networked databases are deliberately designed to be used by researchers and students, rather than librarians. They enable you to do a comprehensive literature search in just a few minutes, with only a minimum of training and practice – or at least, that is the intention. However, nothing is as simple as it looks, and CD-ROM searching is full of traps and pitfalls for the novice. There is a (true) story of the experienced scientist performing a CD-ROM search on his own subject of research, and failing to find even his own papers!

This booklet aims to describe how to set about the task of literature searching. It explains how best to use the major secondary sources – the abstracting and indexing services on CD-ROM. It also describes the printed secondary services, which may still be superior to CD-ROM for some purposes. Finally, in Chapter 5 there is advice on what to do if some of the material you need is not to be found in your library.

Chapter 2. Reference Sources

A science library typically contains five kinds of literature:
- *Books* – ususally referred to as 'textbooks' and 'monographs'. This category may also include conference proceedings.
- *Scientific journals* – which may also include conference proceedings, if they occur regularly, e.g. annually.
- *Theses and dissertations.*
- *'Secondary sources'* – guides to information published in books and journals.
- *'Grey Literature'* – i.e. material not published commercially. This may include specially commissioned reports, pamphlets, training manuals, etc.

Increasingly, libraries also contain video and multimedia material.

The bulk of all scientific information is originally published as articles ('papers') in **journals** and **conference proceedings**. There are two kinds of paper: original reports and reviews. Both contain lists of references for further reading; reviews give longer reference lists than original papers, but tend to be less up-to-date. Accounts of important findings may later be described in **books**, which are often like collections of review articles. Books usually contain long reference lists but the biggest reference lists of all are to be found in **theses** and **dissertations**. These contain detailed accounts of research that is also published (in less detail) in journals, together with much that is not published elsewhere. Few people ever read theses right through!

The reference lists to be found in articles, books and theses can often be the most important part of them, and are certainly the easiest starting point when one has to draw up a reference list of one's own. However, they are not enough by themselves, for various reasons. Firstly, they are never quite up-to-date. An article in a journal generally appears in print between one and two years after it was written, and most of the references it cites will probably have been a year or two old even then. Secondly, reference lists are never entirely comprehensive. They reflect their authors' preoccupations (and areas of ignorance), and as such can never be relied on completely.

So it is best to look in the **secondary sources**. These are publications which index and/or summarise information that has been published in journals or (less often) books. Most of them are now also available on CD-ROM. There are two main kinds: encyclopaedias and dictionaries; and indexing and abstracting services. In this chapter we will look at encyclopaedias and dictionaries.

Encyclopaedias are particularly useful to medical students. They provide essays on specialist subjects for quick reference. They are especially useful in pharmacology.

However, the information in them cannot be relied on to be completely up-to-date. Four of the most important are:

- *The British Pharmacopoeia* – Published by Her Majesty's Stationery Office. Contains monographs on medicinal substances, giving molecular structure, characteristics, identification, assay, action and use (in one or two words). It also has more general mongraphs on: Formulated Preparations; Blood Products; Immunological Products; Radiopharmaceutical Preparations; Surgical Materials. The Appendices describe analytical techniques. The 1993 edition is the most recent; addenda are published annually.

- *Martindale: the Extra Pharmacopoeia*
  This contains entries on all the world's drugs, describing their formulation, uses, side-effects, dosage etc., and giving reference lists. It is now available on CD-ROM as well as in print, but the printed form is probably just as quick and easy to use.

- *The Merck Index*
  This gives a brief description of every drug there is, including its molecular structure.

- *Meyler's Side Effects of Drugs*, and its updates the *Side Effects of Drugs Annual*.
  This contains 51 chapters, each covering a class of drugs. Again, extensive reference lists are provided. The CD-ROM equivalent is *Sedbase*.

Chapter 3. Electronic Searching

The other main secondary sources are the **abstracting** and **indexing services**. These are periodicals which provide guides to current scientific literature: they tell you what has been published, and where. Today, these are generally searched electronically, although they are still published in printed 'hard-copy' format (see next chapter). There are two main methods of electronic searching: **CD-ROM (compact disc read-only memory)** and **online**. Both require a desk-top computer. For CD-ROM, the computer is usually linked to a compact-disc drive, although the CD-ROMs can be networked, so many users can access them remotely. Most university libraries are now linked to the BIDS network, which carries several of the most important databases. For online searching, the computer must be connected to the public telephone network, usually by means of a modem. The two systems are very similar to use, except that CD-ROM is generally intended for the occasional user, while online is aimed at the full-time information scientist. So CD-ROM software is generally more 'user-friendly', but tends to be less versatile than online.

For users at universities, networked CD-ROM databases often have the advantage that each record is marked with a tag indicating whether or not the library holds the journal in question.

CD-ROM searching is an interactive process. The searcher types in a command on the keyboard, and the computer displays the result achieved, giving the number of references it retrieves. The references themselves will only be seen when the searcher gives a 'display' or 'print' command. Usually the commands are provided as a menu, but on some systems the option of a **command language** is available. A language is quicker and more versatile than a menu, but less user-friendly.

The four of the most important abstracting services in life sciences are:

- *Biosis Previews (Biological Abstracts)*
- *Medline (Index Medicus)*
- *Embase (Excerpta Medica)*
- *Science Citation Index*

These are described briefly as follows.

Medline. Published by the U.S. National Library of Medicine. This is the most commonly-used medical secondary service, and probably the most important. It provides an index to journal articles over the whole field of medical and medical-related sciences. For each paper it gives the title, authors, journal name in its standard abbreviated form (your library will have a list of journal names and their abbreviations), volume and page numbers, and date. Most papers also have abstracts attached; the exceptions are letters and short communications, and editorials.

Many people think of *Medline* as being the complete guide to medical journal literature, but this is not so. It only covers a minority of journals, albeit these tend to be the most established and prestigious ones. Its particular weakness is a strong bias towards North American journals. It is also generally a little out-of-date. Journal articles do not appear in it until several months after they have been published. *Medline* can be accessed on CD-ROM or on-line, including via the British Library's BLAISE-LINK/GRATEFUL MED service.

Embase (Excerpta Medica). Published by Elsevier Scientific Publishing. This is the other major medical secondary service. Not every library has it, but if it is available it is wise to use it. The CD-ROM version published by SilverPlatter looks exactly like *Medline*, and one may be tempted to think that there is no point in using it as well. In fact, there are marked differences in coverage: the two databases only overlap by about 30-40%. *Embase* generally gives better coverage than *Medline* in the following areas: new journals; continental European journals; pharmaceutical research. It is also generally kept more up-to-date than *Medline*. For users in university libraries, *Embase* also has the advantage of being available on the BIDS system of networked databases.

Biosis Previews. Published by Bio Sciences Information Service (BIOSIS), Philadelphia. *Biosis Previews* is the main abstracting service covering the whole of life sciences. Its coverage of medicine is not as extensive as *Medline* or *Embase*, but it is good, nonetheless. Its real strength, however, lies in other areas of biology. Like *Medline* and *Embase*, the CD-ROM version is published by SilverPlatter, using the same software as the other two databases.

Basic searching

How do you set about devising a 'search strategy' for compiling a good reference list on a given subject? The important trick is to **break the subject down** into a set of simple concepts, search for each concept separately, and then combine them. This is done using the words **AND**, **OR** and **NOT**. These are the 'Boolean Operators', and are the special feature of all electronic database searching. The key to effective searching is a clear understanding of how they work. Their meanings are best described by using examples. An example of a simple search is given on pages 8-9.

Keyword searching

The advantage of a simple search like the one on pages 8-9 is its speed and simplicity. Its disadvantage is that it is liable to retrieve irrelevant papers along with the relevant ones. In this case, no account was taken of the context in which the words 'propranolol', 'bronchial' and 'smooth muscle' occurred. Several references may only have mentioned propranolol in passing, or could even have been about the effects of propranolol, 'other than in bronchial smooth muscle...' Moreover, it may miss some relevant papers.

To avoid these problems, all the databases have special indexing features, designed to make searching strategies more accurate. Chief among these is the use of 'keywords', added to each reference. An example of a keyword search is given on pages 10–12.

A keyword cannot be any word you like. It has to be one of the permitted terms listed in the database's **thesaurus**. For *Medline*, this is the *Medical Subject Headings* ('*MeSH*') thesaurus. Your library should have a copy. (Copies are available for purchase from the British Library, tel. 01937 546039) *MeSH* includes keywords, and also **sub-headings**. These denote broad subject areas, such as 'pharmacokinetics' or 'veterinary medicine', which can be combined with keywords to make your search more precise.

The keywords themselves are arranged in a 'tree structure', so that you can choose a broad concept for a wide-ranging search, or a narrower concept if you wish to be more precise. E.g. if you look up ASTHMA, you will find that it is listed as coming under the 'broader term' BRONCHIAL DISEASES, which in its turn comes under RESPIRATORY TRACT DISEASES. ASTHMA itself covers two 'narrower terms': EXERCISE-INDUCED ASTHMA and STATUS ASTHMATICUS. N.B. *MeSH* uses American spelling throughout: e.g. FETUS, ANEMIA, SULFUR, etc.

Excerpta Medica has its own thesaurus, called *Emtree*. On the SilverPlatter system it can be reached at the beginning of a search by keying in T⟨Return⟩, which selects the 'Thesaurus' option. There are two types of keywords: Descriptors (DE) and Drug Descriptors (DR). The Drug Descriptors are all the official names (i.e. not the trade-names) of drugs. The menu commands for using the thesaurus are quite self-explanatory.

Biosis Previews has a Search Guide, including a dictionary of keywords. Your library will have copies of these. Unlike *MeSH* and *Emtree,* the keywords are not tree-structured, but this is no great drawback. For more ambitious searchers, *Biosis Previews* has five-figure Concept Codes, e.g. CC175★ is the Concept Code for 'muscle', and CC17506 codes for 'muscle pathology'. It also has Biosystematic Codes for taxa of organisms: BC855★ codes for birds, BC85558 codes for parrots and cockatoos. Unfortunately, neither the keywords nor the Concept and Biosystematic Codes are included on the disc. You have to look them up in the printed Search Guide.

Science Citation Index Published by the Institute of Scientific Information Inc. Bimonthly, with annual cumulations. *Science Citation Index* lists all the papers that have been published in the most important scientific journals, and lists all the papers that each one cites in its reference list. It is especially useful, because if you know a relatively old (i.e. more than two years old) paper of particular interest to you, you can then see which papers have cited it recently. The chances are that at least some of them will be worth reading. Like *Embase*, the *Science Citation Index* is available on the BIDS networked system.

Worked examples of searches in the *Science Citation Index* are given on pages 12–14.

Science Citation Index covers the whole of science, not just biology and medicine. However, it only covers the 3000 most important journals. Sometimes this is probably just as well, because otherwise it would be very cumbersome to use, and would give far too many obscure references which could in practice be safely ignored. However, for an exhaustive literature search, e.g. for a doctoral thesis, this selectivity is a drawback.

The four databases described here are not the only important ones in medical and life sciences. In particular, students of nursing and community health care may need to know of the **Nursing and Allied Health** (*CINAHL*) database, and also possibly **Psycinfo** (*Psychological Abstracts*). These are abstracting services similar to *Medline*, *Embase* and *Biosis Previews*. The **Social Science Citation Index** and the **Arts and Humanities Citation Index** are similar to the *Science Citation Index*, and are available to university libraries on the BIDS network. The **Allied and Alternative Medicine Database** (*AMED*) is a useful source of complementary medicine, occupational therapy, physiotherapy, podiatry and rehabilitation material. Usefully, all items included on *AMED* are available as a photocopy or on loan from the British Library Document Supply Centre.

Disadvantages of CD-ROM

CD-ROM is impressive at first sight, but it has its limitations. To start with, not every secondary service is yet available on it, although the number is increasing fast. CD-ROM is rarely as up-to-date as the printed sources. Few databases are updated at more than three-monthly intervals in their CD-ROM form; the printed version is typically updated monthly. And although a compact disc can store a huge amount of information, most of the larger databases are so big that they are spread over many discs (usually one for each year). Large institutions are increasingly networking their CD-ROM drives, so that several discs can be searched simultaneously, and changing databases can be done by a few keystrokes. But if your CD-ROM is on a stand-alone machine, the job is much slower. Many disc drives still only take one disc at a time, so searching through even one database can mean repeating the same search several times, changing discs each time.

Online databases

The other way to search databases by computer is online. All the secondary sources available on CD-ROM have their online versions. These are generally the same as the CD-ROM databases, although they may go back further in time. The usual method of gaining access is through a '**host**' company, which mounts it on its own mainframe computer, together with software to enable searching, using a command language. There are several hosts. They charge users for the time spent online and for each reference retrieved and printed out. British Telecom also charges for the connect time, so online searching can rapidly become expensive. Connect time charges can however be reduced in some cases by being connected via the Internet.

Advantages and disadvantages of online searching

Advantages:
1. Speed. Online searching is generally quicker than CD-ROM searching. This is mainly because the command languages used are less cumbersome than menu-driven software.

2. Multiple database searching. Most of the major hosts permit one to search through several databases simultaneously. If the same reference turns up in more than one database, a simple command will remove the duplicate versions. It is almost as quick to search through *Medline, Embase* and *Biosis Previews* simultaneously as it is to search through one of them alone.

3. Currency. Online databases are updated much more often than CD-ROMS, typically fortnightly.

Disadvantages:
1. Cost. Online searching is expensive, both in overheads and running costs. In addition to having the necessary hardware, it is also advisable to buy one of the various software packages designed to make searching quick and easy; one must subscribe to one or more hosts, and one must keep up-to-date copies of the manuals and thesauri of the various databases. And then there is the cost incurred in performing each search. Any library performing more than, say, two or three major searches per week will probably find CD-ROM cheaper.

2. Online searching needs technical expertise. Each host has its own command language and range of special features, which have to be learnt. It is best to attend the courses that the hosts provide before trying to do it oneself.

For these reasons, library users are generally not expected to do their own online searching. You have to get the librarian to do it for you.

Basic CD-ROM search: a worked example

We will look for references on the subject '*the effects of Propranolol on bronchial smooth muscle*'. We will be searching in *Medline* discs covering the period from January 1990 to March 1995, using SilverPlatter software. Other systems have a different appearance, but the principles are the same.

When we start a search, the flashing cursor appears at the bottom of the screen next to the word FIND:. We type in the words and phrases we are looking for:

PROPRANOLOL Return
BRONCHIAL Return
SMOOTH MUSCLE Return

Note that we have broken our subject down into its constituent parts. The system does allow us to type in whole phrases like 'effects of propranolol on bronchial smooth muscle', but it is very unwise to try this. If we did, we would miss all the references that did not contain that exact phrase, word for word.

The screen now looks like this:

```
SilverPlatter 3.11                          F10=Commands F1=Help

No.    Records        Request
#1     4969           PROPRANOLOL
#2     10041          BRONCHIAL
#3     23728          SMOOTH
#4     64836          MUSCLE
#5     20127          SMOOTH MUSCLE


Type a search then Enter (↵). To see records use Show (F4). To Print use (F6).
```

This tells us that there are 4969 references in the database mentioning the word 'propranolol', 10041 references mentioning 'bronchial', etc. Each of the 'search statements' on the screen has a number, #1, #2 etc. We will be using these. (The lines labelled #3 and #4 were generated automatically by the computer. We can usually ignore them.)

We want the references which mention all these words together. To do this we use the Boolean Operator **AND**. We key in:

#1 AND #2 AND #5

AND means 'occurring together with', so '#1 AND #2' means 'propranolol' occurring together with 'bronchial'. A new line appears on the screen:

```
#6        35          #1 and #2 and #5
```

This tells us that there are 35 references in the database mentioning the words 'propranolol', 'smooth muscle' and 'bronchial' all together. This is a manageable number.

We can now look at these references simply by pressing the **F4** key, which displays them. If we see a reference of particular interest, we can print it out. In the SilverPlatter system, this is done by moving the cursor to any point in the reference, and pressing **Return**. This 'marks' the reference. Then we can press **P Return**, and that reference is printed out.

Each reference consists of the following 'fields':

TI	–	Title
AU	–	Author(s)
AD	–	Author's address
SO	–	Source (i.e. journal in which the paper was published, with volume, page numbers and date)
ISSN	–	Journal's International Standard Serial Number (probably irrelevant)
PY	–	Year of publication
LA	–	Language of the main text (the abstract is always in English, but the rest of the paper may not be).
CP	–	Country of publication (but not necessarily the country where the authors work).
AB	–	Abstract (some papers do not have this).
MESH	–	Keywords (their use is described below)
TG	–	Checktags (used by advanced searchers; not described in this book)
NM	–	Names of drugs mentioned.
AN	–	Accession Number (probably irrelevant)
UD	–	Update (probably irrelevant)

If we want to get the original paper off the library shelf, we must note down the **SO (Source)** field. This gives the journal's name, and the volume and page numbers of the paper.

Medline Keyword search: a worked example

We will look for references on *'the pharmacology and pharmacokinetics of Propranolol in smooth muscle, other than vascular muscle'*.

Medline on *SilverPlatter* has its *MeSH* thesaurus mounted on the compact disc, so there is no need to get the printed version off the shelf. We reach it by pressing the **F5** key. When we do this, The word FIND at the bottom of the screen is replaced by:

```
INDEX word to look up: _
Type the word or root you want to look up in the INDEX, then ENTER
```

We then type in the first term we are looking for: **PROPRANOLOL** Return. A list of terms and sub-headings appears on the screen:

SilverPlatter 3.11	F10=Commands	F1=Help
Word	Occurrences	Records
PROPRANOLOL	15716	4969
PROPRANOLOL-	31	29
PROPRANOLOL-ADMINISTRATION-AND-DOSAGE	332	332

etc.

The top word on the list is highlighted. We move the highlight down the list with the arrow keys until we come to PROPRANOLOL-PHARMACOKINETICS. We then press Return to select it, and F to 'find' it, i.e. look for references containing it.

We then repeat the process for the other keywords we need, each time pressing F5 to get into the thesaurus; typing in a word; selecting a term by using the up and down arrow keys and the Return; and then pressing **F** to 'find' it. The terms we include in our search in this way are:

PROPRANOLOL-PHARMACOLOGY
MUSCLE-SMOOTH-DRUG-EFFECTS
MUSCLE-SMOOTH-VASCULAR-DRUG-EFFECTS

After we have searched for all these terms, the screen looks like this:

```
SilverPlatter 3.11                              F10=Commands F1=Help

No.      Records     Request
#1        138        PROPRANOLOL-PHARMACOKINETICS
#2        1947       PROPRANOLOL-PHARMACOLOGY
#3        3157       MUSCLE-SMOOTH-DRUG-EFFECTS
#4        4242       MUSCLE-SMOOTH-VASCULAR-DRUG-EFFECTS
```

We now combine these terms using the Boolean Operators **AND**, **OR**, and **NOT**. We type in **#1 OR #2**. This gets us #5, containing everything indexed either under PROPRANOLOL-PHARMACOKINETICS or PROPRANOLOL-PHARMACOLOGY, or both.

Next, we type in **#3 NOT #4**. This gets us #6, containing all the references indexed under MUSCLE-SMOOTH, but not MUSCLE-SMOOTH-VASCULAR.

Finally, we type in **#5 AND #6**. The screen now looks like this:

```
SilverPlatter 3.11                              F10=Commands F1=Help

No.      Records     Request
#1        138        PROPRANOLOL-PHARMACOKINETICS
#2        1947       PROPRANOLOL-PHARMACOLOGY
#3        3157       MUSCLE-SMOOTH-DRUG-EFFECTS
#4        4242       MUSCLE-SMOOTH-VASCULAR-DRUG-EFFECTS!
#5        2059       #1 or #2
#6        2990       #3 not #4
#7         69        #5 and #6
```

This tells us that we have retrieved 69 references that fit our profile. We can now start displaying them on the screen by pressing **F4** in the normal way.

Searching in *Science Citation Index* online: worked examples

These examples show the *Science Citation Index* as it appears with its own proprietary software. The version available on the BIDS network looks different, but the principles of searching are the same.

1. Simple author search

We will search for papers by the author A. Arak (a zoologist), published in 1994, and see what papers they cite, using a computer with a single-disc drive. The introductory screen looks like this:

```
---- V3.05 ------ Science Citation Index (Jan 94 - Dec 94) -- 3.0 -------------
F1 - Help    F2 - Database    F3 - Search    F4 - Results    F5 - Quit
-------------------------------------- Search Session ------------------------
Set   Records   Field
        1       Title
                Enter as single words or phrases: CELL or INTERLEUKIN 2
                Press ↵ to execute search
                ┌──────────────────────────────────────────────────┐
                │ _                                                │
                └──────────────────────────────────────────────────┘

        Alt-Fields  Alt-Dictionary   Alt-Limit   Alt-Undo   Alt-CopyQuery
        Alt-ClearSession  Alt-PrintSession  Alt-SaveStrategy  Alt-RunStrategy
```

We start by pressing **Alt** and **F** to change from title-searching to author-searching; a list of Fields (i.e. paragraph headings) appears. Using the arrow keys we move the highlight down to **Author**, then we press **Return**.

Now we enter the author's name: **ARAK-A*** **Return**. The asterisk is very important: it is a 'truncation' or wild-card symbol; if we leave it off, we will not retrieve anything. The screen now looks like this:

```
---- V3.05 ------ Science Citation Index (Jan 94 - Dec 94) -- 3.0 -------------
F1 - Help    F2 - Database    F3 - Search    F4 - Results    F5 - Quit
-------------------------------------- Search Session ------------------------
Set   Records   Field
 1       1      Author
                ARAK-A*
```

13

This tells us we have retrieved one paper. To see the citation, we press **F4**, then **Return**. This gives us a new screen:

```
------------------------------ Records: 1 of 1 ------------------------------

Enquist-M  Arak-A
Symmetry, Beauty and Evolution   (English)   => Article
NATURE
Vol 372    Iss 6502    pp 169-172 1994 (PQ688)

Related Records: 20        Cited References: 25
```

This is a reference to an article in *Nature*, which cited twenty-five references. We can display these references by pressing **F**.

2. Citation search
We will search for papers published in 1994, citing the classic paper published by J. K. Waage (1979) in *Behavioral Ecology and Sociobiology*, volume 6, page 147.

To start with, we press **Alt** and **F**, as in the previous example. We then move the highlight to Citation. From there on, we perform the search in exactly the same way as an author search.

We enter **WAAGE-JK-1979-BEHAV-ECOL-SOCIOBIOL-V6-P147**. We press **Return**, and the screen looks like this:

```
----- V3.05 ------ Science Citation Index (Jan 94 - Dec 94) -- 3.0 -------------

F1 - Help    F2 - Database    F3 - Search    F4 - Results    F5 - Quit
---------------------------------------- Search Session -----------------------------------
Set    Records   Field
1         3      Citation
                 WAAGE-JK-1979-BEHAV-ECOL-SOCIOBIOL-V6-P147
```

This tells us that three papers published in 1994 cited Waage's 1979 paper. If we press the **F4** key the first of these will appear. We press the **Page Down** key to see the next, and **F** to see all the papers it cites (including Waage's).

If we do not know the full bibliographic details of Waage's original paper, we can use the asterisk truncation mark:
WAAGE-JK-1979* will retrieve all papers citing anything Waage published in 1979, **WAAGE-JK*** will retrieve all papers citing anything by Waage in any year.

Chapter 4. Manual Searching

When CD-ROMs are easily available, one might wonder why anyone should want to use the old-fashioned printed versions of the secondary sources. In fact, there are times when it is at least as quick and easy. Anyone using a crowded library, at a college for example, where people have to wait for their turn on the CD-ROMs, should seriously consider using the printed versions instead. They particularly come into their own for answering quick questions, such as when you know of an article, but are not sure which journal published it, or who all the authors were, or the exact title. Generally, if you only want to look for references published in one particular year, the printed secondary services are often as quick to use as the CD-ROMs, if not quicker.

This chapter describes the printed 'hard-copy' versions of the secondary services described in the last chapter, and also *Current Contents*.

Current Contents. Published by the Institute for Scientific Information. This is the simplest of the secondary services. It consists simply of the Contents pages of current editions of scientific journals. It is published weekly, which makes it the most up-to-the-minute of all the secondary services. For obvious reasons, it is divided into a number of different subject areas, all published separately. The ones of interest to the biologist are: Clinical Medicine; Clinical Practice; Agriculture, Biology and Environmental Sciences; Agriculture, Food and Veterinary Sciences; Life Sciences. As a tool for literature searching, its use is limited, because it is not indexed at all. However, there are times when it comes in handy. For instance, if you have heard that a particular author has recently published a paper, and you cannot get hold of the journal in question (either because your library does not subscribe to it, or because the latest issue has not arrived yet), *Current Contents* will give you the article's full title, full list of co-authors, and page numbers. If you want to keep your reading up-to-date, but you cannot spend more than a minute or two in the library, *Current Contents* will at least tell you what you ought to be reading. But it is not geared for comprehensive searching. *Current Contents* is also available online and on diskette as *Current Contents Search*. University users can access it on the BIDS network.

Index Medicus. Published by the US National Library of Medicine. The printed version of *Medline*, but with the bibliographic details of papers only, without the abstracts. Monthly, with annual cumulations. *Index Medicus* is published in three parts: Author Index, Subject Index, and Bibliography of Medical Reviews.

> The **Author Index** is self-explanatory. If you know the name of the author you want to read, you look in the Author Index to see what he/she has published.

The **Subject Index** is useful for browsing in. It has two levels of indexing: headings and sub-headings. These are the same as the Descriptors and Sub-Headings used in CD-ROM searching, and listed in the *MeSH Thesaurus*. Thus, for example, under the heading SPINAL CORD INJURIES, you find the sub-headings Blood, Cerebrospinal Fluid, Classification, Complications, etc. etc. References are indexed under all the appropriate headings ; e.g. a paper entitled *Methicillin-resistant Staphylococcus aureus: an increasing threat in New Zealand Hospitals* (Martin, D. R. et al, N Z Med. J., 1989, Jul 26, vol.102 (872) pp.367-369) is to be found indexed under the headings METHICILLIN (subheading Pharmacology), and again under STAPHYLOCOCCUS-AUREUS (subheading Drug Effects), to name but two.

The **Bibliography of Medical Reviews**, which is also a subject index, is especially useful if you want to compile a short(-ish) reading list quickly. Use it to find the most recent review article on your subject, then pull that review down off the shelf and note down titles from its list of references.

The advantages of *Index Medicus* are its size and scope, and its comprehensive subject index. The disadvantages are a distinct American bias, and the lack of abstracts. **Index Medicus** is available for purchase from the British Library (tel. 01937 546039).

Excerpta Medica. Published by Elsevier Scientific Publishing. The printed version of *Embase*. University libraries rarely subscribe to it these days, but industrial laboratories may still take it. It is arranged quite differently from *Index Medicus*: it consists of a series of 47 sections on different medical topics. Each section is published as a separate journal (monthly, with annual cumulations). They list references to original papers, giving title, authors, authors' address, journal, date, volume and page numbers, and an abstract. Each issue contains a table of contents, giving broad subject headings, a subject index (very detailed), and an author index.

Excerpta Medica can be used to search for papers on a given subject in the following way.

1. Choose which Section is relevant to your interest, e.g. Radiology (Section 14).
2. If your interests are relatively broad, look on the contents page of the latest issues (or annual cumulations), and choose the most appropriate heading, e.g. 4.1., Interventional Radiology.
3. If your interests are narrower, look for them in the subject index, e.g. Radiotherapy. Each reference has a number (usually four figures), the numbers of each reference indexed under Radiotherapy are given, together with all the other words it is indexed under. (This is very useful; it tells you which reference is most strictly relevent to your needs.) Thus under Radiotherapy, you may be most interested by the paper which also lists the keywords Desmoid tumor, Fibromatosis, Surgery, Adult, Histology.

Biological Abstracts. Published by Bio Sciences Information Service (BIOSIS), Philadelphia. The printed version of *Biosis Previews. Biological Abstracts* can be searched in several ways. There is an **Author Index**, giving surnames and initials; this is all right if your author's name is fairly distinctive, but less helpful if it is a really common one like 'K. Suzuki'. There are also three different kinds of subject index:

> **The Biosystematic Index** lists references under the taxonomic Order of the subject organism, e.g. *Marsupialia, Macropodidae* (= kangaroos). All medical papers appear under Primates, Hominidae. Under these biosystematic headings are given subject sub-headings: e.g. under Primates, Hominidae, there are such sub-headings as Addiction, Adrenals, Allergy, etc., etc. But this still leaves one with an enormous number of references to choose from if the subject is cancer therapy, for instance. So its use is limited.
>
> **The Generic Index** lists references against the species they mention, together with one subject concept: e.g. *Drosophila melanogaster.* Genet. Animal (= animal genetics). This is probably more useful than the Biosystematic Index for non-human biology, but it does not include papers about *Homo sapiens* at all.
>
> **The Subject Index** lists references under each of their key words: each keyword is placed in the center of a 60-character string taken from the complete set of keywords to that reference, e.g. a paper may be indexed under '… lial cell proliferation ACELLULARITY capillary dilation micr …': this paper will also be indexed under '… ration endothelial cell PROLIFERATION acellularity capillary …', etc. This is intended to make subject-searching easy, but in practice it can be time-consuming, and it is easy to overlook a reference in all the columns of small print. The best way to scan through it is to cover the page with a sheet of paper, which you then move down the page. As each line of print appears from under it, glance at it to see if it contains the keywords you are looking for.

Science Citation Index. Published by the Institute for Scientific Information. The printed version is published bimonthly with annual accumulations. *Science Citation Index* is published in three parts.

1. The **Source Index**, giving all the papers covered in that edition, listed alphabetically by their first author. Each entry gives: First author's name, co-authors, title, source (i.e. journal, volume and page numbers, date), and the first author's address. So for example, if you want to see if D. R. Jacobson has published anything in the last year, you look up JACOBSON DR: under this name, the Source Index gives the full details of two papers.

2. The **Citation Index** itself, giving all the authors cited by papers listed in the Source Index, again listed alphabetically, e.g. if you want to see if anyone has cited the works of R. L. Trivers (an eminent sociobiologist) during the last year,

you look in the Citation Index. Here, under TRIVERS RL you will find a list of all his papers that have been cited, and all the papers that have cited them.

3. The **Permuterm Subject Index**. This lists the papers given in the Source Index, under the words in their titles. E.g. if you want to see what papers have been published in the last year with 'noradrenergic denervation' in the title, you can look in the *Permuterm Subject Index* under the heading NORADRENERGIC, and find that it lists DENERVATION under that heading, with one author's name against it: EISON AS (you then look this author up in the Source Index); or you can look under DENERVATION, and find the same author's name against the word NORADRENERGIC.

A major drawback of the *Science Citation Index* is that it only refers to papers under the first author's name.

Chapter 5. If the Library Doesn't Have It...

We have now seen how we can compile a reading list of journal papers on a given topic in the life sciences, using either manual or computerised methods. The next step, of course, is to get hold of the papers themselves. This is no problem if the library has the journals in question. But what if it hasn't? The traditional answer used to be to ask for a reprint from the paper's main author. However, this practice is now dying out; fewer and fewer journals are providing their authors with a supply of reprints.

So if your own library does not have a particular reference, you must look for a library that does. This is quite easy in principle. Many libraries have a 'union list' – a catalogue of the stock of several other libraries. Before visiting the library of another institution, you should find out about its admissions policy. Generally a National Union of Students membership card or similar identification is sufficient to get you in, but some libraries have more restrictive policies.

But first, you must find which libraries hold the publication you need. This can be done online, using **JANET – the Joint Academic Network**. This provides access to the computerised library catalogues of most of Britain's universities and other further education institutions, and also other laboratories funded by the research councils. The easiest way to get access to these is probably via **BUBL** – the Bulletin Board for British Libraries.

The **Internet** and the **World Wide Web** are becoming useful for literature searching, as more and more institutions are producing their own Web pages. At the time of writing, the South and West Health Care Libraries have a very comprehensive Web page, which includes library users' guides. The address is **http://cochrane.epi.bris.ac.uk/rd/links/libs.htm**. Other NHS regions will doubtless soon follow suit. However, the Internet is notorious for the variable quality of the information found on it, and for the difficulty in tracking anything down. The **OMNI** project (**Organising Medical Networked Information**), based at the National Institute of Medical Research, Mill Hill, aims to provide a single site for medical information on the WWW. The address is **http://omni.ac.uk**. It has the advantage of being selective; all information is evaluated before being included on the site. Information is indexed, and most importantly, descriptions are provided of the sources the information comes from. At present OMNI is concentrating heavily on information from sources in the United Kingdom.

Finally, there is the *BRITISH LIBRARY*. Everyone has heard of the Round Reading Room in the British Museum; less well known are the **Science Reference & Information Service (SRIS)** and the **Document Supply Centre (BLDSC)**.

SRIS is Britain's national science library. It subscribes to some 29,000 current journals, as well as having an excellent collection of secondary sources, books, and the world's largest collection of patents (including patents for pharmaceuticals, medical equipment and genetically engineered organisms). At present this material is based in two main sites in central London: biomedical literature is held at the Aldwych Reading Room at 9 Kean St. (off Drury Lane), together with literature on earth sciences, astronomy and mathematics; patents, literature on the physical sciences, engineering and business are at the Holborn Reading Room at 25 Southampton Buildings (off Chancery Lane). SRIS is open to the general public: no reader's ticket or identification is needed to gain admission. The Aldwych Reading Room is open from 9.30 a.m. to 5.30 p.m. on weekdays. In addition to its collection of books, journals and all the printed 'hard-copy' secondary sources described in this book, it also has a CD-ROM facility available for public use, with *Medline*, *Excerpta Medica (Embase)*, *Biosis Previews* and several other databases. (It is advisable to book in advance if you want to use these. Tel. 0171-412-7288.) The Holborn Reading Room also has *Science Citation Index* on CD-ROM. SRIS also has the 'STM search' service, which will perform online searches for a fee (tel. 0171-412-7477). This service also has access to many more databases not described here.

New from SRIS, is the **Health Care Information Service**, (tel. 0171-412-7933) which produces the *AMED* database of literature citations and abstracts on allied and alternative medicine (available as hardcopy extracts, on floppy disk and online via the host Data-Star), the *Medline Update* series of current-awareness bulletins on various medical topics and on-line access to the National Library of Medicine's suite of databases including *MEDLINE* and *TOXNET via BLAISE-LINK/GRATEFUL MED*.

SRIS is a reference library: it does not lend any books or journals, although it does have a photocopy service. If you wish to obtain a copy of a paper from the British Library without visiting SRIS, you should ask your library to obtain one from the **Document Supply Centre** (tel. 01937 546060). The BLDSC provides a document delivery service to organisations and individuals throughout the world. There are various requesting methods: mail, fax, e-mail, via certain database hosts, and on-line to the BLDSC's automated request processing system. Requests are normally dealt with within 48hours of receipt and are dispatched by first class post. Where greater speed is essential the Urgent Action Service will respond to your request in less than two hours and can fax back the results. At the time of writing the basic cost of a DSC request to registered UK customers was just over £4.00 per item.

Further information about SRIS and the Document Supply Centre may also be obtained from: Marketing & Public Relations Section, Science Reference & Information Service, 25 Southampton Buildings, London WC2A 1AW. Alternatively, a complete guide to the British Library's services may be found at its WWW home page 'PORTICO', address **http://portico.bl.uk/**.